Copyright © 2020 by Littlestone

All rights reserved. No part of this publication may be reproduced, distributed or transmitted in any form or by any means, without prior written permission.

Special Operations Fitness - Hell Week / Paperback -- 1st ed.
ISBN: 978-1-946373-10-6
$34.99 everywhere great books are sold

SPECIAL OPERATIONS FITNESS – HELL WEEK

Table of Contents

This Fitness Program is for You…..6
Disclaimer 6
Background 6
Special Operations Fitness Workout Principles 8
- ★ Train as You Fight: 8
- ★ Individual Discipline and Motivation: 8
- ★ Teamwork or with a Battle Buddy: 9
- ★ Don't Overtrain: 9
- ★ Self-Defense: 9
- ★ Proper Running Technique: 10
- ★ You Don't Need Expensive Gear or Equipment: 10
- ★ Discipline is Essential 10
- ★ Diet is Essential: 10
- ★ Sleep is Essential: 10
- ★ Dress Appropriately: 11
- ★ Long, Fast Walks: 12
- ★ Warm Up Before. Only Stretch After: 12
- ★ Plan Ahead: 12
- ★ Safety: 12

Special Operations Fitness - Hell Week 13
Workout Descriptions 14
   Syria 14
   Iran 14
   Egypt 14
   Tidal Wave 15
   Afghanistan 15
   Somalia 16
   Iraq 18
   Israel 19
   Self-Defense 1 19
   Self-Defense 2 19
   Self-Defense 3 20
   Self-Defense 4 20
   Self-Defense 5 21
   Self-Defense 6 21
Training Variations for Aspiring Navy SEALS 23
Conclusion 23

SPECIAL OPERATIONS FITNESS – HELL WEEK

THIS FITNESS PROGRAM IS FOR YOU…

… if you are interested in pursuing a career in Special Operations and want to know if you have what it takes to make it through training.

… if you are already a strong and capable athlete looking to push yourself though an unconventional cross-training program.

… if you like to challenge yourself to accomplish great things.

… if you want a "gut check."

… if you want to prove that you still got it.

DISCLAIMER

Beginners should not attempt Special Operations Fitness – Hell Week.

Only those men and women who are already in great shape should attempt Special Operations Fitness – Hell Week.

Seek medical approval prior to doing Special Operations Fitness – Hell Week.

Conduct Special Operations Fitness – Hell Week at your own risk.

BACKGROUND

Greetings, my name is Christopher Littlestone and I am a retired US Army Special Forces (Green Beret) Lieutenant Colonel, Airborne Ranger & Combat Diver. Having been to and through some of the hardest military training programs in the world, I get asked a lot of questions from my YouTube and social media network about "how strong do I need to be to make it through Special Operations Training?"

The answer is never simple. Every school, selection, and training event is different.

Let me give you three quick examples:

<u>Ranger School</u>: Ranger school is 20–24 hours of activities each day. The harassment factor (aggressively in your face) is constantly a 9 out of 10. The easiest days are 20 hours of

LIFE IS A SPECIAL OPERATION.COM

practical exercises, combatives, obstacle courses, and physically demanding training. You get two meals and a couple hours of sleep. The hardest days are when you are patrolling 24 hours straight. You get two Meals Ready to Eat (MREs) and zero sleep.

Special Forces Selection: I will be vague and say that the most sleep I ever got during selection was four hours. But that was the exception. There were also days when we only got 15 minutes of sleep. Some of those days we would do physical training (PT), go for a timed long-distance run, go for a timed long-distance ruck march, and then prepare our gear for the next day. Bedtime was 0230, only to be awoken at 0245 for morning log PT and another full day of physically demanding events.

The Special Forces Combat Diver Qualification Course: "Dive School" has four workouts a day. You do crushing PT in the morning. Then you go to breakfast. After breakfast, you have three hours of intense training and exercising in the pool. After lunch, you have an open water swim or dive. After dinner, you have another dive in the dark. You get to sleep as much as you can, eat as much as you want, but at the end of the day, you are exhausted from four world-class workouts. Did I mention that on the last day of training, we did a nine-mile formation run? Fallouts didn't graduate. For both SF Selection and Dive School, there is a very professional and intense tone. But for the most part, there is less harassment than Ranger School. But when it comes, it comes hard – 10 out of 10.

Special Operations Training - Are You Ready for It?

SPECIAL OPERATIONS FITNESS – HELL WEEK

The point is this: I've seen and done training on all sides of the spectrum. Whether it is from harassment or from stringent standards, all of them had one thing in common: *they demanded and evaluated physical fitness and mental toughness, hours of it each day.*

The Stealth Marathon: A few years after finishing my Special Forces training, I was on leave, visiting family in the Rocky Mountains. I had been there several times before and had even enjoyed some Boy Scout backpacking trips there when I was in high school. My third day in town, I woke up early and arrived at my favorite trail head at 5:00 a.m. with a small water bottle and a power bar. I then ran four hours uphill, to the top of the snow-topped peak. It took me three hours to run down. I took a quick shower, then met everyone at our favorite Mexican restaurant for a late lunch. Physical fitness is a way of life for those within the Special Operations community. We don't train for an entire year to finish a four-hour marathon. We do seven hours of hard trail running just like that … because we are fit and hard and nothing scares us. And then we go for lunch like nothing happened.

I developed Special Operations Fitness – Hell Week to give you a taste of what the Special Operations community is like.

Remember …

This workout is easier than the easiest week of Ranger School
This workout is easier than the easiest week of SCUBA School
This workout is easier than the easiest week of Special Forces Training
This workout is easier than the easiest week of Navy SEAL Training
This workout is easier than the easiest week of MARSOC Raider Training
This workout is easier than the easiest week of Air Force Special Ops Training

I think you get the point. Do this workout to get a glimpse of what you are going to need to be able to give in order to endure the rigorous training required to make it into the elite Special Operations community.

SPECIAL OPERATIONS FITNESS WORKOUT PRINCIPLES

★ <u>Train as You Fight:</u> Special Operations Fitness – Hell Week is going to help you build muscle, strength, health, and confidence so you are fully prepared for the challenges of your life. Train like someone is watching you and evaluating to see if you give 100% or if you are a slacker.

★ <u>Individual Discipline and Motivation:</u> Without individual discipline, there is no way to be the best of the best. Inspiration and motivation must come from within. Special

Operations Fitness – Hell Week is designed to smoke you, to challenge you. No one is yelling at you or pushing you. This is because those who are truly inspired to be the best of the best don't need external motivation.

★ Teamwork or with a Battle Buddy: All workouts have been designed for the individual. But feel free to work out with a "Fitness Buddy" or in a small group to minimize the risk of injury or accident. Although working in a team is always a good way to keep you accountable and motivated, you are not competing with or against a team. You are competing against yourself in an attempt to be stronger, smarter, and more physically fit than you have ever been.

★ Don't Overtrain: I've warned future military athletes countless times not to overtrain. Clearly, Special Operations Fitness – Hell Week is overtraining. So please mitigate the beating that your body is going to take by allowing it an appropriate amount of time to recover after the week is over. I would not recommend doing Special Operations Fitness – Hell Week just before shipping out to Special Operations training. Give yourself a few weeks to recover. And please, enhance your recovery through proper sleep and diet.

JTACs embedded with Army Conventional Forces

★ Self-Defense: Special Operations Fitness – Hell Week includes eight self-defense / cardio kickboxing sessions. These self-defense workouts focus on basic boxing and kicking techniques and are guaranteed to "wear you out." Whether an experienced warrior or an amateur soccer mom, you will benefit from these challenging and confidence-building workouts.

SPECIAL OPERATIONS FITNESS – HELL WEEK

★ <u>Proper Running Technique:</u> Using proper running technique increases speed and cardio-vascular efficiency, reduces muscular-skeletal impact, and decreases the probability of injury. I wish every kid had to take track-and-field "sprinting" workshops in school. Once you learned how to run properly, you instantly feel and see the difference. It is never too late to learn proper form: Don't heel strike. Land on your mid or forefoot. Swing your arms front and back, not sideways. Head up. Knees high. Torso erect, slightly forward-leaning. Fast cadence. Use your hamstrings … Please learn how to run properly. Your joints will thank you.

★ <u>You Don't Need Expensive Gear or Equipment:</u> Special Operations Fitness – Hell Week minimizes the use of equipment. We require that you have a punching bag (heavy bag), hand wraps, punching-bag gloves, a pullup bar, a couple of dumbbells, a backpack, and access to a place where you can swim. You can buy these tools for lifelong use, use equipment at a gym or a friend's house, or make some of them yourself. Be flexible and adapt.

★ <u>Discipline is Essential</u>: To be the best of the best, discipline and unwavering resolve are always required. Thousands of people waste money paying personal trainers each week to hold their hand from fitness machine to machine, even though they already know how the machine works. This is because they lack discipline. Please "Ranger Up" and find, make, or build the discipline required to finish this life-changing Special Operations Fitness – Hell Week program.

★ <u>Diet is Essential:</u> To be a world class performer, you need to be fueled by real (not processed) food. No one puts cheap gas into a race car. When I was a young Special Forces Officer, I lived off of cheeseburgers, protein shakes, and Starbuck's Frappuccinos. I survived, but in spite of my bad eating habits. Once I learned more about the importance of eating healthy, and started to actually eat an abundance of real, healthy food, my performance increased significantly. Please learn from my sophomoric mistakes and feed your body healthy, real food. Now is not the time to go on a starvation diet. If you are going to use this fitness program as a prerequisite for attending Special Operations training in the future, then rest assured, you will be given an opportunity to do work in a significant caloric deficit. But for the sake of minimizing muscular-skeletal fatigue and maximizing strength and endurance, please eat several thousands of healthy calories during Hell Week.

★ <u>Sleep is Essential:</u> Most people are sleep deprived and functioning below their potential. Now add into your normal day-three workouts, a swim, and a ruck march. I guarantee that you are going to be exhausted. I don't limit your sleep during Special Operations Fitness – Hell Week. This is because I want you to get all the benefits of world class

LIFE IS A SPECIAL OPERATION.COM

training without enduring the mind-numbing effects of sleep deprivation. So please maximize your sleep and recovery each night.

★ <u>Dress Appropriately:</u> We recommend wearing the appropriate protective clothing when exercising. Usually this means boots and a military-style uniform. Wear running shoes when doing the track workouts to minimize the chance of injury. Swim in pants and a t-shirt. Psychologists have realized that when a person puts on a uniform, they become the part. So wear a Special Operations Fitness uniform and you will become more intense, more focused, and it will make things more fun. "Mandex" (Man Spandex) and "Wandex" (Woman Spandex) are unauthorized. If you are doing Special Operations Fitness – Hell Week to prepare for future Special Operations training, then you have no option… wear your uniform. Combat Divers swim and dive in full uniform. PJs go on rescue missions in uniform. MARSOC Raiders and Rangers ruck all night while patrolling in uniforms. Train as you fight.

Helicopter Casting ("Helo-Cast") into a lake for a training mission.
How far can you swim in full uniform?

SPECIAL OPERATIONS FITNESS – HELL WEEK

★ <u>Long, Fast Walks:</u> Special Forces soldiers and Infantry men might walk (ruck) all day. Literally, they might have their rucksacks (backpacks) on for 24 hours. Don't forget that an M4 rifle weighs seven pounds or that a M240B machine gun weighs 27 pounds. Not only is long-distance walking a significant part of Special Operations training, it can be a low-impact exercise beneficial for people of all ages. When you do a long, fast walk, you burn calories. But even after you finish a long, fast walk, you continue to burn calories. This is why you can go on a backpacking trip, eat chocolate bars all day, and come back skinnier and leaner than when you left. Carrying a heavy backpack and body armor is a part of Special Operations, one which frequently causes life-long disabilities in the neck and back. For this reason, the recommended ruck weight for Special Operations Fitness – Hell Week is 35 pounds. Make sure that you carry food, water, a flashlight, a first-aid kit, and your cell phone. The added physical stress of carrying a weapon will be simulated by carrying a small seven-pound dumbbell. You will shred away calories, burn fat, build core body muscles, get fresh air, and enjoy hours of low-impact, high-benefit walking.

★ <u>Warm Up Before. Only Stretch After:</u> It is important to get your heart pumping and muscles warmed up before starting your workout. A few minutes of biking, jogging, or jumping jacks should do the trick. Never stretch a cold muscle. Stretching a cold muscle is an easy way to tear or injure that muscle. Only stretch after your muscles are warm, but preferably after the workout as part of your cool-down. Please warm up before every workout and only stretch after the workout.

★ <u>Plan Ahead:</u> World class training requires an efficient use of time. For example, every hour of every day of Ranger School is planned for Ranger students and printed on the master training plan (of course the students don't get to see the plan). Special Operations Fitness – Hell Week has made a very detailed fitness plan for you to execute during a normal work or school week. Please take a few minutes before you start to plan your workouts and routes ahead of time. Plan what and when to eat. Plan when to sleep and when to wake up. Plan where to go to have the required dumbbells and how to get access to a swimming pool and a heavy bag. Be deliberate and plan ahead so you can successfully complete each day.

★ <u>Safety:</u> Special Operators have dangerous jobs. They rely on good safety habits to stay alive. They wear helmets, gloves, and eye protection; use seatbelts in helicopters; and do rigorous SCUBA gear and parachute inspections. You need to be smart about protecting yourself. Consult a Doctor before doing Special Operations Fitness – Hell Week. Please remain hydrated and well fed. Always let someone know where you are going hiking, or simply bring a "Fitness Buddy." Swim in the presence of a lifeguard. Don't get run over because you are so focused on your music, podcast, or audio book. Take one earbud out so you can maintain situational awareness, and listen for approaching traffic or threats.

SPECIAL OPERATIONS FITNESS: HELL WEEK

HELL WEEK

Day #	1 Saturday	2 Sunday	3 Monday	4 Tuesday	5 Wednesday	6 Thursday	7 Friday	8 Saturday
Before Breakfast Morning Workouts	Syria	Iran	Egypt	Somalia	Syria	Iran	Egypt	Somalia
	Self-Defense 1	Self-Defense 2	Self-Defense 3	Self-Defense 4	Self-Defense 5	Self-Defense 6	Self-Defense 4	Self-Defense 5
	Tidal Wave	Tidal Wave	Tidal Wave	Tidal Wave	Tidal Wave	Tidal Wave	Tidal Wave	Tidal Wave
9-5 Day Job			8-Hour Workday	8-Hour Workday	8-Hour Workday	8-Hour Workday	8-Hour Workday	
Daytime	Afghanistan	Afghanistan	Afghanistan	Afghanistan	Afghanistan	Afghanistan	Afghanistan	Afghanistan
After Work	Iraq	Iraq	Iraq	Iraq	Iraq	Iraq	Iraq	Israel

SPECIAL OPERATIONS FITNESS – HELL WEEK

WORKOUT DESCRIPTIONS

Syria

- ★ Concept:
  - 100m Walking Lunges
  - 1 Mile Run for Time
  - 100m Walking Lunges
- ★ Equipment Needed: None. I recommend you do this workout at a track.
- ★ Note: Lunges are not timed, so don't rush and please don't use sloppy technique. Take larger than normal steps when doing your lunges and never let your knee go forward of toes. Warm up before officially beginning the workout. Stretch immediately after finishing the workout. Practice proper running technique.

Iran

- ★ Concept: At full effort, sprint 800m. Rest 2 minutes. Repeat.
  - Do 6 sets of 800m for a total of 4,800m or 3 miles.
- ★ Equipment Needed: None. I recommend you do this workout at a track.
- ★ Note: Warm up before officially beginning the workout. Stretch immediately after finishing the workout. Practice proper running technique.

Egypt

- ★ Concept: At full intensity, conduct the following hillside / sprint workout.
  - Sprint 3 minutes uphill, walk or jog back to the starting position
  - Run backwards 3 minutes uphill, walk or jog back to the starting position
  - Run sideways (grapevine) 3 minutes uphill, walk or jog back to the starting position
  - Skip 3 minutes uphill, walk or jog back to the starting position
  - Side-shuffle 3 minutes uphill, walk or jog back to the starting position
  - Sprint 3 minutes uphill, walk or jog back to the starting position

- ★ Equipment Needed: +/- 600m long hill or road.
- ★ Note: Warm up before officially beginning the workout. Stretch immediately after finishing the workout. When running sideways or doing the side-shuffling, alternate sides every 50 meters.

Tidal Wave

- Concept: Swim 1000 meters.
- Equipment Needed: Pants, belt, t-shirt at a minimum. Wear googles or a mask if you prefer.
- Note: You can do whatever stroke you want. If you are potential Navy SEAL candidate, then you must do 2000m, not just 1000m, each day using the combat side stroke. Warm up before officially beginning the workout. Stretch immediately after finishing the workout.

Surface Swimming (Side Stroke) in the Pacific Ocean

Afghanistan

- Concept: During your normal workday, take breaks every hour or so to do some simple calisthenics. By the end of your workday, you must complete the following:
    - 100 Pushups
    - 100 Leg Lifts
    - 50 Pullups
- Equipment Needed: A pullup bar and 2 even surfaces from which to do your leg lifts.
- Note: Build muscle memory via perfect technique. Pause (briefly) at the top of each Pushup, Pullup, and Leg Lift. I recommend 5 breaks during your workday where you do 10 Pullups, 20 Pushups, & 20 Leg lifts.

Leg Lifts (Use a chair, desk … whatever is available)

SPECIAL OPERATIONS FITNESS – HELL WEEK

Somalia

- ★ Concept: At full intensity (but with control so you don't injure yourself), do 20 minutes of Traveling Man Makers with very light dumbbells. Record how far you travel.
- ★ Equipment Needed: 10- or 20-pound dumbbells, and a safe road or track.

Note: Warm up before officially beginning the workout. Stretch immediately after finishing the workout. Do perfect form exercises to reduce the chance of injuries.

LIFE IS A SPECIAL OPERATION.COM

"Traveling Man Maker" explained (from left to right, top to bottom): Walking Lunge right leg. Walking Lunge left leg. Squat down, thrust legs back into pushup position (plank). One Pushup. Right Arm Dumbbell Row while in Left Arm Plank. Left Arm Dumbbell Row while in Right Arm Plank. Legs in. Stand up. Bicep Curls to Military Press. Dumbbells back to side. Repeat.

SPECIAL OPERATIONS FITNESS – HELL WEEK

<u>Iraq</u>

- ★ Concept: Carrying a rucksack (backpack) of 35 pounds, hike (don't run) as fast as you can for 3 hours. Carry a 7-pound weight in one hand. This will simulate an M4 rifle.
- ★ Equipment Needed: A 35-pound backpack and a 7-pound dumbbell. If you are a potential member of Special Operations, then I insist that you do this workout in your uniform and military boots. "Train as You Fight."
- ★ Note: Do not stash away your dumbbell or carry it in your backpack. Alternate hands as you want, but always carry your dumbbell. Warm up before officially beginning the workout. Eat or drink all you want while hiking. Maintain your "Situational Awareness" and don't get hit by cars or bike riders. Stretch immediately after finishing the workout. The minimum military standard for a ruck march is four miles per hour. So in three hours on a road or trail or sidewalk, you should be able to walk twelve miles. If you are walking cross-country on uneven terrain, don't worry about distance. Just go as far and as fast as you can.

MARSOC Raiders during training. How far can you carry your gear, a weapon, and a casualty?

LIFE IS A SPECIAL OPERATION.COM

Israel

- Concept: Beginning 3 hours before sunset and carrying a rucksack (backpack) of 35 pounds, hike (don't run) as fast as you can for 6 hours. Carry a 7-pound weight in one hand. This will simulate an M4 rifle.
- Equipment Needed: A 35-pound backpack and a 7-pound dumbbell. If you are a potential member of Special Operations then I insist that you do this workout in your uniform and military boots. "Train as You Fight."
- Note: Begin this workout 3 hours before sunset. This means you will do 3 hours of hiking before sunset and 3 hours of hiking after sunset. Bring a flashlight and reflective belt as required. Do not stash away your dumbbell or carry it in your backpack. Eat or drink all you want while hiking. Maintain your "Situational Awareness" and don't get hit by cars or bike riders. Alternate hands as you want, but always carry your dumbbell. Warm up before officially beginning the workout. Stretch immediately after finishing the workout. Take a shower, eat something healthy, drink some electrolytes, and get some sleep.

Self-Defense 1

- Concept: At full intensity (but with control so you don't injure yourself), do the following circuit 5 times.
    - 50 left hand jabs
    - 50 right hand jabs
    - 50 left hand punches
    - 50 right hand punches
    - 50 hooks / upper cuts (1 left then 1 right) 100 total
    - 2 minutes rest
- Equipment Needed: Hand wraps, bag gloves, heavy bag.
- Note: Warm up before officially beginning the workout. Stretch immediately after finishing the workout. Do perfect form strikes. Never do a sloppy strike. If you get tired, then slow down. Build muscle memory via perfect technique.

Self-Defense 2

- Concept: At full intensity (but with control so you don't injure yourself), do the following circuit 5 times:
    - 50 repetitions of 2 left hand jabs followed by 1 right hand punch
    - 50 repetitions of 2 right hand jabs followed by 1 left hand punch
    - 50 repetitions of 2 left hand jabs followed by 1 right hand punch, then a left hook

SPECIAL OPERATIONS FITNESS – HELL WEEK

- - 50 repetitions of 2 right hand jabs followed by 1 left hand punch, then a right hook
  - 2 minutes rest
- ★ Equipment Needed: Hand wraps, bag gloves, heavy bag.
- ★ Note: Warm up before officially beginning the workout. Stretch immediately after finishing the workout. Do perfect form strikes. Never do a sloppy strike. If you get tired, then slow down. Build muscle memory via perfect technique.

Self-Defense 3

- ★ Concept: At full intensity (but with control so you don't injure yourself), spar with the heavy bag for 3 minutes then rest for 2 minutes. Use only punching techniques. Repeat 6 times.
- ★ Equipment Needed: Hand wraps, bag gloves, heavy bag.
- ★ Note: Warm up before officially beginning the workout. Stretch immediately after finishing the workout. Do perfect form strikes. Never do a sloppy strike. If you get tired, then slow down. Build muscle memory via perfect technique.

Self-Defense 4

- ★ Concept: At full intensity (but with control so you don't injure yourself) do the following circuit 5 times.
  - 20 right rear leg front kicks
  - 20 left rear leg front kicks
  - 20 right front leg front kicks
  - 20 left front leg front kicks
  - 20 right rear leg round kicks
  - 20 left rear leg round kicks
  - 20 right front leg round kicks
  - 20 left front leg round kicks
  - 2 minutes rest
- ★ Equipment Needed: Heavy bag.
- ★ Note: Warm up before officially beginning the workout. Stretch immediately after finishing the workout. Keep your hands up to protect your face. Do perfect form kicks. Never do a sloppy kick. If you get tired, then slow down. Build muscle memory via perfect technique.

LIFE IS A SPECIAL OPERATION.COM

Self-Defense 5

- ★ Concept: At full intensity (but with control so you don't injure yourself), spar with the heavy bag for 3 minutes, then rest for 2 minutes. Use punches and kicks. Repeat 6 times.
- ★ Equipment Needed: Hand wraps, bag gloves, heavy bag.
- ★ Note: Warm up before officially beginning the workout. Stretch immediately after finishing the workout. Do perfect form punches and kicks. Never do a sloppy punch or kick. If you get tired, then slow down. Build muscle memory via perfect technique.

A Ranger on a mission in Iraq. How heavy is his gear, body armor, M249 machine gun & ammo?

Self-Defense 6

- ★ Concept: At full intensity - conduct the following 6 heavy bag circuits.

    ○ Round 1:
        ▪ 50 left hand jabs
        ▪ 50 right hand jabs

SPECIAL OPERATIONS FITNESS – HELL WEEK

- 50 left hand punches
- 50 right hand punches
- 50 hooks or upper cuts (1 left then 1 right) 100 total
- 2 minutes rest

○ Round 2:
- 20 right rear leg front kicks
- 20 left rear leg front kicks
- 20 right front leg front kicks
- 20 left front leg front kicks
- 20 right rear leg round kicks
- 20 left rear leg round kicks
- 20 right front leg round kicks
- 20 left front leg round kicks
- 2 minutes rest

○ Round 3:
- 50 left hand jabs
- 50 right hand jabs
- 50 left hand punches
- 50 right hand punches
- 50 hooks (1 left then 1 right) 100 total
- 2 minutes rest

○ Round 4:
- 20 right rear leg front kicks
- 20 left rear leg front kicks
- 20 right front leg front kicks
- 20 left front leg front kicks
- 20 right rear leg round kicks
- 20 left rear leg round kicks
- 20 right front leg round kicks
- 20 left front leg round kicks
- 2 minutes rest

○ Round 5: 3-minute spar with heavy bag using kicks and punches. Rest 2 minutes.

○ Round 6: 3-minute spar with heavy bag using kicks and punches.

LIFE IS A SPECIAL OPERATION.COM

TRAINING VARIATIONS FOR ASPIRING NAVY SEALS

If you are an aspiring Navy SEAL, add 1000 meters to each day's swim. This will make 2000m each day. That's 20 more minutes of swimming each day. Deal with it! Use the Combat Side Stroke.

CONCLUSION

Special Operations Fitness – Hell Week
- ★ +/- 100 Miles of Ruck Marching
- ★ 5 Miles of Swimming (SEAL Variant includes 10 miles)
- ★ 800 Pushups
- ★ 400 Pullups
- ★ 6x Track or Hill Sprint Workouts
- ★ 2x Cross-Training Sessions
- ★ 8x Cardio Kickbox Workouts

If you are doing the Special Operations Fitness – Hell Week as a life-changing challenge, then I wish you well. Congratulations on being courageous enough to undertake such a difficult test.

If you are doing the Special Operations Fitness – Hell Week to prepare for future military and Special Operations training, Good luck. I know this will prepare you for the rigorous trials you will face and achieve. Please keep me posted as you progress through the various phases of your training.

Now … get to work!

Respectfully,

Christopher Littlestone
Life is a Special Operation
Are You Ready For It?

SPECIAL OPERATIONS FITNESS

HELL WEEK

by

Life is a Special Operation.com